Light and Sound
A Range of Energy Waves

by Rebecca L. Johnson

Table of Contents

Develop Language 2

CHAPTER 1   Light Makes It Bright 4
                    Your Turn: Observe 9
CHAPTER 2   Visible and Invisible Waves 10
                    Your Turn: Interpret Data 15
CHAPTER 3   A World of Sound 16
                    Your Turn: Summarize 19

Career Explorations 20
Use Language to Explain 21
Science Around You 22
Key Words 23
Index 24

DEVELOP LANGUAGE

Three…two…one. Lift off!

You see the bright **light** at the bottom of the rocket's engines. Then, seconds later, you hear the roar of the rocket's engines. The rocket lifts off and carries a satellite into space.

The satellite will take photos and send them back to scientists on Earth.

Discuss the photos on these pages with questions like these:

You see the light from the rocket before you hear its sound. Why do you think that happens?

The rocket's engines make, or produce, the sound. What do you think produces the light coming from the bottom of the rocket?

What do you notice about the lights in the satellite photo?

light – energy that travels in transverse waves

satellite

Light and Sound: A Range of Energy Waves

CHAPTER 1

Light Makes It Bright

Light from the sun and stars travels to Earth through space. Light travels in **transverse waves** at about 300,000 kilometers (186,000 miles) per second. It takes about 8 minutes for light to travel from the sun to Earth.

Light from distant stars may take hundreds, thousands, or even millions of years to reach Earth.

transverse waves – waves whose energy moves forward and whose crests and troughs move up and down

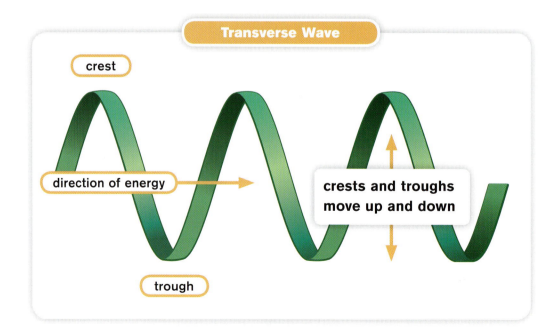

4 *Light and Sound: A Range of Energy Waves*

Light Wave Properties

There are many kinds of light waves. Each kind of light wave can be described by its properties.

For example, **wavelength** is a property of waves. Wavelength is the distance from one wave crest to the next or from one wave trough to the next. **Frequency** and **amplitude** are also properties of waves. They describe how much energy a wave has. High-frequency waves have more energy than low-frequency waves. High-amplitude waves have more energy than low-amplitude waves.

wavelength – the distance from one wave crest to the next or from one wave trough to the next

frequency – the number of wavelengths that pass a point in a certain length, or unit, of time

amplitude – the distance from the middle of a wave to a crest or a trough

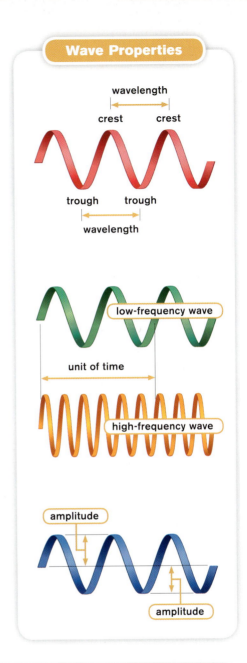

KEY IDEAS Light waves are transverse waves. Wavelength, frequency, and amplitude are properties of light waves.

Chapter 1: Light Makes It Bright 5

How Light Waves Act

Light waves travels in straight lines until they hit a surface. When light waves hit a surface, three things can happen. The light waves may be **absorbed** by the surface. They may bounce off the surface. Or they can pass through the surface.

Most objects don't give off their own light. Instead, the light waves coming from a source bounce off the object. The bouncing of light waves off an object is **reflection**. You see these objects when the reflected light waves strike your eyes.

A mirror reflects almost all the light that strikes it and forms an **image**. The light waves that form the image have been reflected twice. First, light waves coming from a light source reflect off a surface or object. Then those reflected light waves strike the mirror and are reflected. When these reflected waves strike our eyes, we see the image in the mirror.

absorbed – taken in
reflection – the bouncing back of light waves off an object
image – a likeness of something

SHARE IDEAS What other objects or surfaces do you know that reflect light like a mirror? **Explain**.

◂ The bird sees its image because light reflected from the bird is reflected again by the mirror.

6 Light and Sound: A Range of Energy Waves

Refraction is the bending of light. Refraction takes place when light waves pass through two different **transparent** materials.

For example, when light travels from air into a block of clear plastic or glass, light waves slow down and change direction slightly. In other words, the light waves refract. Refraction can make some things look very strange!

refraction – the bending of energy waves

transparent – allows light to pass through

Explore Language
LATIN WORD ROOTS
transparent
trans (through) + *parere* (to appear) = to appear through

KEY IDEA When light waves strike a surface, they may be absorbed, reflected, or refracted.

Light waves that pass from the air into the plastic block are refracted.

Chapter 1: Light Makes It Bright

Light and Lenses

Refraction can also make something look closer and clearer. A lens is a piece of transparent glass or plastic that is shaped to refract light waves.

People use telescopes to study stars and other objects in space. Some telescopes have lenses. Some have mirrors. Other telescopes have both lenses and mirrors. The lenses and mirrors refract and reflect light so that distant objects can be seen.

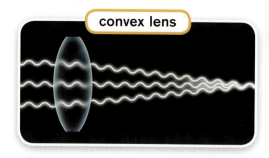

convex lens

▲ A convex lens brings light waves closer together. A convex lens can make an object appear larger than it really is.

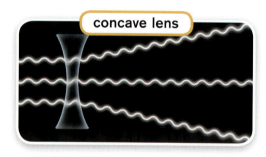

concave lens

▲ A concave lens spreads light waves apart. A concave lens can make the detail in an object clearer.

Keck Telescope

KEY IDEA A lens refracts light waves to form an image.

8   *Light and Sound: A Range of Energy Waves*

YOUR TURN

OBSERVE

Look at this photo or at images reflected in a mirror. Make a list or a sketch of everything you see.

Share your list with a friend. Then answer the following questions.

- What was the source of the light waves that formed the images in the mirror?

- How were the light waves reflected before they reached your eyes?

MAKE CONNECTIONS

Telescopes have lenses that make distant objects look closer and clearer. Microscopes have lenses, too. What do microscopes do?

USE THE LANGUAGE OF SCIENCE

What is the difference between reflection and refraction of light waves?

In reflection, light waves bounce off an object. In refraction, light waves bend.

Chapter 1: Light Makes It Bright 9

CHAPTER 2
Visible and Invisible Waves

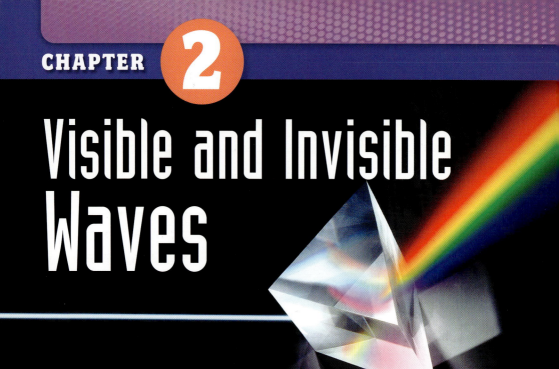

Energy we can see is **visible light**. The light may be reflected light, or it may come directly from a light source such as the sun. Sunlight looks white. But when white light strikes a glass **prism**, the light leaves the prism as a **color spectrum**.

The band of colors forms because white light refracts as it passes through the prism. The light waves in white light have many different wavelengths.

Each color in the spectrum is made up of light with certain wavelengths. We see the colors because some waves bend a little more, or a little less, as they pass through the prism.

visible light – energy waves that human eyes can see

prism – a piece of clear glass that refracts visible light

color spectrum – a band of light made up of red, orange, yellow, green, blue, and violet light

10 *Light and Sound: A Range of Energy Waves*

Objects around us are many different colors. The color of an object depends on how it absorbs and reflects light. A red apple, for example, absorbs all the colors of light except red. The apple reflects red light, so we see the apple as red.

▲ We see an apple as red because it reflects red light.

▶ The sun produces more than visible light waves.

The sun and stars produce more than just visible light waves. They produce waves that have the same properties as light waves, but are **invisible** to our eyes. We can't see these energy waves.

Like visible light waves, invisible waves can travel through empty space. They also travel at the same speed as visible light.

However, invisible waves have wavelengths and frequencies different from visible light. Some have more energy than visible light waves. Some have less.

invisible – cannot be seen by human eyes

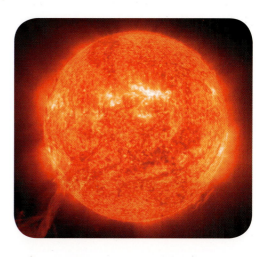

Chapter 2: Visible and Invisible Waves 11

The Electromagnetic Spectrum

◀ X-rays easily pass through skin, but not bones. Doctors use these high energy waves to make X-ray images of bones.

X-rays
0.00000003 mm to 0.000003 mm

⬅ shorter wavelengths, higher energy

gamma rays
0.000000003 mm to 0.00000003 mm

Gamma rays are extremely powerful. They have the most energy of all the waves in the electromagnetic spectrum. Gamma rays are used to fight some types of cancer.

ultraviolet rays
0.00003 mm to 0.0003 mm

Humans can't see ultraviolet rays. However, some insects and other animals can. Banks use ultraviolet rays to identify fake paper money.

The range of energy waves that includes all visible and invisible light waves is called the **electromagnetic spectrum**.

electromagnetic spectrum – the complete range of visible and invisible energy waves

Light and Sound: A Range of Energy Waves

▶ Microwaves are the energy waves that heat food in microwave ovens. The radar that tracks weather also uses microwaves.

visible light

microwaves

0.3 cm to 300 cm

longer wavelengths, lower energy

infrared waves

0.0003 cm to 0.03 cm

radio waves

3 m to 300 m

Humans can't see infrared waves. When infrared waves are absorbed, their energy is changed to heat. Many restaurants use infrared waves to keep food hot.

Radio waves carry television and radio programs through space. Some carry cell phone and wireless Internet signals.

0.0004 mm to 0.0007 mm

Waves in the electromagnetic spectrum are arranged by the amount of energy they have.

KEY IDEA The electromagnetic spectrum is made of visible and invisible energy waves.

Chapter 2: Visible and Invisible Waves   13

Seeing Invisible Waves

Scientists use instruments, or tools, to **detect** invisible energy waves. For example, radio telescopes on Earth detect radio waves coming from objects in space, such as the sun.

Some instruments that detect invisible energy waves are on satellites far above Earth. These instruments may detect X-rays, infrared light, ultraviolet light, or microwaves.

Computers often turn the information collected by these instruments into pictures we can see.

detect – see or sense

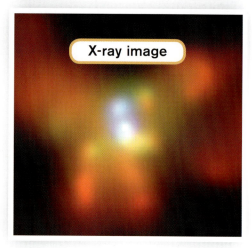

▲ Instruments on satellites detect different types of energy waves.

KEY IDEA Scientists use instruments to detect invisible light energy waves.

14 *Light and Sound: A Range of Energy Waves*

YOUR TURN

INTERPRET DATA

Look at the chart. Interpret the data to answer the questions. Share your answers with a friend.

1. Which color has waves with the shortest wavelength?
2. Which color has waves with the lowest frequency?
3. Which color's waves carry the most energy?

Colors of Visible Light	Wavelength in millimeters
violet	0.00038–0.00045
blue	0.00045–0.00049
green	0.00049–0.00057
yellow	0.00057–0.00059
orange	0.00059–0.00062
red	0.00062–0.00075

MAKE CONNECTIONS

The ultraviolet rays in sunlight can hurt your skin. Two kinds of products can protect your skin. Sunscreens absorb ultraviolet rays. Sunblocks reflect these rays. Draw a diagram to show how these products are different. Tell when you should use these products.

STRATEGY FOCUS

Make Connections

Make connections to the information in this chapter. Talk about other things you have read about or experienced that are related to these ideas about energy waves.

Chapter 2: Visible and Invisible Waves

CHAPTER 3

A World of Sound

We can hear music on a radio. The music is carried through the air by radio waves we cannot hear or see. The radio turns radio waves into sounds our ears can hear.

Sound travels through matter as compression **waves**. Sound waves are made when particles of matter vibrate, or move quickly back and forth. These **vibrations** travel as waves through matter.

This is one way in which light waves and sound waves are different. Light waves can travel through empty space, but sound waves cannot.

compression waves – waves whose energy moves forward and whose particles move back and forth

sound – energy that travels in compression waves through matter

vibrations – quick back-and-forth movements

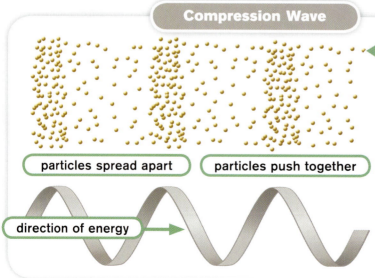

In a compression wave, particles of matter alternately push together and spread apart. The pushed-together areas are like crests. The spread-apart areas are like troughs.

16 *Light and Sound: A Range of Energy Waves*

Like light waves, sound waves can be described by properties such as wavelength, amplitude, and frequency. A sound's **volume** is related to its amplitude. High-amplitude sounds are loud, while low-amplitude sounds are soft.

A sound's **pitch** is related to its frequency. High-frequency sounds have a high pitch, while low-frequency sounds have a low pitch.

Like light waves, sound waves can be reflected and refracted. A reflected sound is an **echo**.

Sound waves travel at different speeds through different kinds of matter. For example, sound waves travel more than four times faster through water than they do through air.

volume – how loud or how soft a sound is

pitch – how high or how low a sound is

echo – a reflected sound

By The Way...

Sound waves can travel very long distances through water. Scientists think that some low-frequency sounds made by whales can travel several thousand kilometers through the ocean.

KEY IDEAS Sound waves are compression waves. Wavelength, amplitude, and frequency are properties of sound waves.

Chapter 3: A World of Sound

Sounds We Cannot Hear

Just as there are light waves we cannot see, there are sound waves we cannot hear.

A sound wave that has a frequency too low for our ears is an **infrasound**. A sound wave with a frequency too high for our ears is an **ultrasound**.

Many bats produce high-frequency, or ultrasound, waves. These bats listen for the echoes the waves make.

When the waves reflect from objects such as buildings and trees, echoes help the bats avoid flying into them. When the waves reflect from flying insects, the echoes help the bats find food.

infrasound – sound waves with a frequency too low for human ears to hear

ultrasound – sound waves with a frequency too high for human ears to hear

▶ Bats have very large ears to help them hear the echoes of the sounds they make.

Infrasound	Ultrasound
frequency too low for human ears	frequency too high for human ears
lower frequency	higher frequency
longer wavelength	shorter wavelength

KEY IDEA There are some high-frequency sounds and some low-frequency sounds that human ears cannot hear.

Light and Sound: A Range of Energy Waves

YOUR TURN

SUMMARIZE

Summarize the information in this chapter by completing these sentences.

1. Sound waves travel through _____ .

2. The properties of sound waves include _____ .

3. We cannot hear infrasounds and ultrasounds because _____ .

MAKE CONNECTIONS

Have you ever noticed a dog perk up its ears and listen to something you didn't hear? Suggest why dogs can hear sounds that people cannot.

EXPAND VOCABULARY

A prefix at the beginning of a word can give clues about the word's meaning.

Prefix	Meaning
infra-	below, under, or within
micro-	very small
ultra-	beyond the range of, or beyond what is ordinary

Find words with these prefixes in this book. Make a list of other words with the same prefixes. Draw pictures and write sentences to show the meanings of these words.

Chapter 3: A World of Sound

CAREER EXPLORATIONS

Satellite Engineers: What Do They Do?

Hundreds of satellites orbit Earth. Some satellites take pictures of Earth and other objects in space. Some keep track of the weather. Other satellites can track the location of planes, ships, trucks, and some cars. There are even satellites that help your cell phone call reach the other side of the world.

Satellite engineers design and build satellites. They also design, build, and test the instruments that satellites carry into space. Satellite engineers spend much of their time solving problems and trying out new ideas.

Many satellite engineers have college degrees. Others have several years of technical training. All of them have a special interest in space and in making tools to help explore it.

- Would you like to be a satellite engineer? Explain.

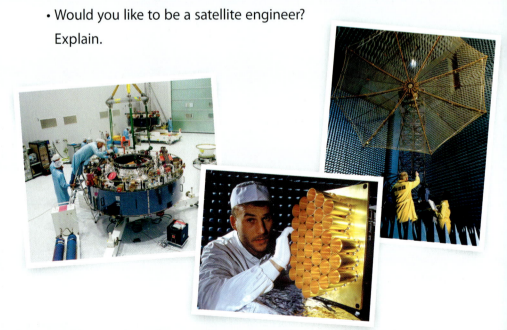

20 *Light and Sound: A Range of Energy Waves*

USE LANGUAGE TO EXPLAIN

Cause and Effect Words

Sometimes you need to explain why something happens. You can give the **causes**, or the reasons, for something. You may also need to explain the **effects** of something by telling what happens. Words such as **so** and **because** can help you explain causes and effects.

EXAMPLES

The grass reflects green light, **so** we see the grass as green.

      cause   effect

We do not see X-rays **because** their wavelengths are too short.

    effect   cause

Look at the photographs on pages 7–8. Tell a friend what is happening to the light waves in these photos. Explain why this is happening to the light waves. Use **so** and **because**.

Write an Explanation

Look at the illustration on pages 12–13. Choose one kind of energy and find out more about it. Explain why this energy is useful.

- Describe the properties of this kind of energy.
- Tell how people use this kind of energy.
- Explain why this kind of energy is useful.

| Words You Can Use ||
Light Wave Words	Cause and Effect Words
frequency wavelength amplitude visible invisible	because so since then

Use Language to Explain

SCIENCE AROUND YOU

Our Singing Sun

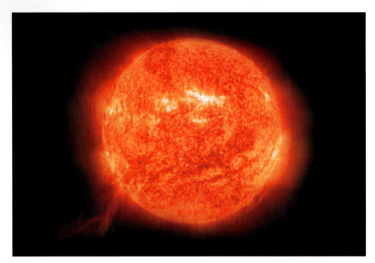

Scientists have discovered something new about our nearest star, the sun. It sings! Satellites that orbit the sun have detected sound waves on the sun. These sound waves come from huge explosions that happen near the sun's surface. However, the sound waves made by the sun are very low-frequency waves. Even if they could travel to Earth, they would be too low for human ears to hear.

Read the article. Then answer the questions below.

- What have scientists recently discovered about the sun?

- How are sound waves on the sun produced?

- Sound waves travel through matter. Why don't sound waves from the sun reach Earth? How is this different from the sun's light waves?

Light and Sound: A Range of Energy Waves

Key Words

compression wave a wave whose energy moves forward and whose particles move back and forth
Sound waves are **compression waves** that human ears can hear.

electromagnetic spectrum the complete range of visible and invisible energy waves
The **electromagnetic spectrum** includes gamma rays, x-rays, microwaves, and radio waves as well as visible light.

gamma rays energy waves with the shortest wavelengths and greatest energy in the electromagnetic spectrum
Gamma rays can be used to keep some food from spoiling.

infrared waves invisible energy waves with wavelengths longer than visible light
A campfire is a source of both visible light and **infrared** waves.

infrasound sound waves with a frequency too low for human ears to hear
Elephants produce **infrasound** that other elephants can hear.

radio waves energy waves with the longest wavelengths and least energy in the electromagnetic spectrum
Some **radio waves** have wavelengths hundreds of kilometers long.

reflection the bouncing back of energy waves when they hit a surface
A large empty room can be a good place to hear echoes, the result of sound **reflections**.

refraction the bending of an energy wave
Refraction causes a band of colors to appear when white light passes through a prism.

transverse wave a wave whose energy moves forward and whose crests and troughs move up and down
Light waves are **transverse waves**.

ultrasound sound waves with a frequency too high for human ears to hear
Doctors use **ultrasound** to "see" inside parts of the human body.

ultraviolet rays waves with wavelengths shorter than visible light
Ultraviolet rays can damage a person's skin.

visible light light waves that human eyes can see
Visible light is made up of several different colors of light.

Key Words 23

Index

amplitude 5, 17, 21
compression wave 16, 17
concave 8
convex 8
crest 5, 16
echo 17, 18
electromagnetic spectrum 12–13
energy 2, 4, 5, 7, 10, 11, 12–13, 14, 15, 16, 21
frequency 5, 15, 17, 18, 21
gamma rays 12

image 6, 8, 9
infrared waves 13, 14
infrasound 18, 19
light 2, 4–8, 9, 10
matter 16–17, 22
microwaves 13, 14
pitch 17
prism 10
property 5, 11, 17, 19
radio waves 13, 14, 16
reflection 6, 9
refraction 7, 8, 9
satellite 2, 14, 20, 22
sound 2, 16–18, 19

spectrum 10, 12, 13
transparent 7
transverse wave 4–5
trough 5, 16
ultrasound 18, 19
ultraviolet rays 12, 14, 15
vibration 16
visible light 10–11, 12, 13, 14, 15
volume 17
wavelength 5, 10, 11, 15, 17, 18, 21
x-rays 12, 13

MILLMARK EDUCATION CORPORATION
Ericka Markman, President and CEO; Karen Peratt, VP, Editorial Director; Lisa Bingen, VP, Marketing; Rachel L. Moir, Director, Operations and Production; Shelby Alinsky, Assistant Editor; Ernestine Giesecke, Science Editor; Pictures Unlimited, Photo Research

PROGRAM AUTHORS
Mary Hawley; Program Author, Instructional Design
Kate Boehm Jerome; Program Author, Science

BOOK DESIGN Steve Curtis Design

CONTENT REVIEWER
Karen Kolehmainen, PhD, California State University, San Bernardino, CA

PROGRAM ADVISORS
Scott K. Baker, PhD, Pacific Institutes for Research, Eugene, OR
Carla C. Johnson, EdD, University of Toledo, Toledo, OH
Donna Ogle, EdD, National-Louis University, Chicago, IL
Betty Ansin Smallwood, PhD, Center for Applied Linguistics, Washington, DC
Gail Thompson, PhD, Claremont Graduate University, Claremont, CA
Emma Violand-Sánchez, EdD, Arlington Public Schools, Arlington, VA (retired)

TECHNOLOGY
Arleen Nakama, Project Manager
Audio CDs: Heartworks International, Inc.
CD-ROMs: Cannery Agency

PHOTO CREDITS cover ©Eddie Gerald/Alamy; IFC and 15b ©David Safanda/iStockphoto; 1 ©Mehau Kulyk/Photoresearchers; 2-3 ©Tom Rogers/Reuters/Corbis; 2 ©Max Dannenbaum/Getty Images; 3a, 3b, 14, 23 ©Getty Images; 4, 5, 12-13, 16b illustrations by Steve Curtis Design; 6 ©William Leaman/Alamy; 7 ©sciencephotos/Alamy; 8a, 8b, 16a illustrations by Joel and Sharon Harris; 8c ©Roger Ressmeyer/Corbis; 9a ©Juniors Bildarchiv/Alamy; 9b ©STILLFX/Shutterstock; 9c and 9d Ken Karp for Millmark Education; 10 ©Clayton J. Price/Corbis; 11a ©D. Hurst/Alamy; 11b ©NASA/epa/Corbis; 12 ©Visuals Unlimited/Corbis; 13 ©Emin Ozkan/Shutterstock; 15a ©Stephen Coburn; 17 ©Kelvin Aitken/agefotostock; 18 ©Dr. Merlin D. Tuttle/Bat Conservation International/Photo Researchers, Inc.; 19 ©Eric Isselêe/Shutterstock; 20a ©Starsem/Photo Researchers, Inc; 20b and 20c ©PHOTOTAKE Inc./Alamy; 22 ©NASA/epa/Corbis; 24 ©Virginia P. Weinland/Photo Researchers, Inc.

Copyright © 2008 Millmark Education Corporation

All rights reserved. Reproduction of the whole or any part of the contents without written permission from the publisher is prohibited. Millmark Education and ConceptLinks are registered trademarks of Millmark Education Corporation.

Published by Millmark Education Corporation
7272 Wisconsin Avenue, Suite 300
Bethesda, MD 20814

ISBN-13: 978-1-4334-0158-9

Printed in the USA

10 9 8 7 6 5 4 3 2 1

Light and Sound: A Range of Energy Waves